Forestry Commission Bulletin 77

GW00363341

British Softwoods:
Properties and Uses

T. Harding
Building Research Establishment,
Princes Risborough Laboratory

LONDON: HER MAJESTY'S STATIONERY OFFICE

ISBN 0 11 710262 8

ODC 8

Keywords: Conifers, Forestry, Wood utilisation

Enquiries relating to this publication should be
addressed to the Technical Publications Officer,
Forestry Commission, Forest Research Station,
Alice Holt Lodge, Wrecclesham, Farnham,
Surrey GU10 4LH

Technical enquiries should be
addressed to the Timber Division,
Building Research Station, Garston,
Watford WD2 7JR

Front Cover *Machine stress graded British grown spruce used as joists in a suspended floor complying with British Standard 5268: Part 2. This house in Middlesex was constructed to the housebuilder's specification by Berkley Homes (North London) Limited, Bushey, Herts. (38264)*

Background *Machine stress graded softwood treated with preservative used to construct a suspended ground floor on a building site at Kettleston Mains in Scotland by Balfour Beatty Limited.* (Forestry Commission, HQ Photographic Library)

Contents

Foreword

During the 1960s and early 1970s, the Princes Risborough Laboratory published a series of monographs on the wood properties of British softwoods. Since that time our knowledge has much advanced, particularly in gaining a better understanding of the features which influence strength properties. Much of the research work was carried out at the Princes Risborough Laboratory and was sponsored by the Forestry Commission. These studies have led to major changes in the grading and supply of British softwoods, and have enabled them to be integrated in the strength class system of BS 5268 : Part 2, which brings together structural timber of all origins in a common classification.

This has also been a period during which the commercial importance of domestic softwood production has grown enormously. Thus in 1987, some 5.0 million cubic metres of softwood were produced and marketed from British forests, more than three times the level of the mid-1960s. This growth in supply has provided the essential raw material for a steady expansion of the British sawmilling and processing industry, and has shown that British forest products can compete successfully in a market which is still largely dominated by imports from Scandinavia, Eastern Europe and North America.

This new publication brings together the latest information on the major British commercial softwood species, and seeks to provide an introduction to conifer wood properties for the guidance of architects, designers, specifiers, foresters and students. I believe that this is a major contribution to our knowledge of British softwoods and I commend it to all who are involved in the use of wood.

G.J. Francis
Director General
Forestry Commission

British Softwoods: Properties and Uses

T. Harding

Building Research Establishment,
Princes Risborough Laboratory

Introduction

During the period 1964–1972, the Princes Risborough Laboratory of the Building Research Establishment issued a series of booklets on the properties and uses of the major British softwoods – Scots pine, Corsican pine, lodgepole pine, Sitka and Norway spruce, Douglas fir and larch (Japanese, European and hybrid) (Forest Products Research Laboratory, 1964; 1965; 1967a; 1967b; 1968; 1972). Since their publication, considerably more information has become available, mostly from the results of studies undertaken within the joint programme of research on British timber, sponsored by the Forestry Commission and carried out at the Princes Risborough Laboratory. These studies covered a wide range of subjects, including the suitability of the major species for use in construction and the effects of forest management practices on timber properties.

The commercial and economic importance of softwoods can be judged from the fact that they account for about 90 per cent of all of the timber used in Britain. Most of the wood used in house building and other construction, in fencing, packaging, pit props, railway sleepers, telegraph and power transmission poles, paper-making and the manufacture of board materials such as chipboard, hardboard and fibre building board, comes from coniferous forests. The bulk of our present-day requirements is imported from the traditional softwood supply areas of North America, Scandinavia and other parts of northern and eastern Europe. Over recent years, however, British softwoods have been claiming a gradually-increasing share of many of the market sectors. This trend is certain to gather momentum over the next two decades as:

1. increasing supplies of well-managed timber become available,
2. there is a greater awareness of the properties and uses of British-grown softwood timber, and
3. more attention is paid to good presentation, particularly of sawn products.

The standards and specifications covering many of the traditional and potential uses of British timber have changed over recent years, in some cases substantially so. This Bulletin provides the information for specifiers and users to make maximum use of the increasing British resource. Its main purpose is to establish the link between requirements for current and potential end-uses and the properties and performance of these commercially important timbers.

A guide to the properties of the individual species is also given for those readers whose main interest is in the timber itself, or in comparisons between species, rather than in specific uses.

The British Softwood Resource

The British softwood resource is almost entirely man-made and has mostly been planted since the 1930s. Though many species are grown, commercial planting is largely of nine species providing four timber types, spruce, pine, Douglas fir and larch, and it is these that are considered in this Bulletin. All, with the exception of Scots pine, have been introduced to British forestry, though for some, such as Norway spruce and European larch, this was as long ago as the 16th and 17th centuries. Characteristically, softwood plantation species grow vigorously and, after 40–50 years, when many of the trees are felled in Britain, they are of comparable size, or larger, than older but more slowly-grown Scandinavian trees.

Sitka spruce

Sitka spruce (*Picea sitchensis*) is the most commonly planted tree in British forests. It is a species of the western seaboard of North America, from southern Alaska to northern California, but most of the trees planted in Britain are of seed of Queen Charlotte Island origin, a provenance combining acceptable vigour and frost hardiness in British conditions. Sitka spruce is favoured because it is readily established, grows well,

Plate 1. *The timber resource in Great Britain is represented here by a mature Scots pine stand (E6841) and its departure from the forest as cross-cut logs* (Inset; E8060).

withstands exposure, is generally disease resistant, has a good stem form and gives a higher yield than most other species, particularly on less fertile sites. It is tolerant of a range of site conditions but grows best in the higher rainfall areas of south and west Scotland, north and south-west England and Wales, and it is from these regions that most of the commercial timber is obtained.

Norway spruce

Norway spruce (*Picea abies*), though not as commonly planted as Sitka spruce, is an important species in British forestry, with a timber supply exceeded only by those of Sitka spruce and Scots pine. It has a natural range over much of northern and central Europe and is better adapted than Sitka spruce for growth in the drier, eastern parts of Britain. It is rarely as vigorous as well-grown Sitka spruce but is of good stem form and modest branch size.

Scots pine

Scots pine (*Pinus sylvestris*) occurs widely through Europe and Russian Asia and, throughout its range, it is an important timber tree. In Britain, it thrives on light and sandy soils but growth is relatively slow compared with that of the spruces. It is frost hardy and can be grown almost anywhere but is most successful in the east; commercial supplies come mainly from central and north-east Scotland, East Anglia and south-east England.

Corsican pine

Corsican pine (*Pinus nigra* var. *maritima*) is one of a number of varieties of the southern European black pine. In Britain, its best growth is on sandy soils in areas of low summer rainfall and correspondingly high summer temperatures, particularly in the midlands and south of England and on the north-east coast of Scotland, notably at Culbin. Where soil and climatic conditions are suitable, Corsican pine produces higher yields than Scots pine due largely to its more vigorous growth in the years following establishment. It is of good stem form, often better than Scots, but has somewhat larger branches in more conspicuous whorls.

Lodgepole pine

Lodgepole pine (*Pinus contorta*) is of comparatively recent use in commercial British forestry. It is a western North American tree with many provenances differing in vigour, form and branching habit. It grows on poor heaths and peats and, being tolerant to exposure, is used as a pioneer species for planting, particularly in western and northern Scotland. Timber volumes are, as yet, small and it will be some years before there is a significant contribution to the market.

Douglas fir

Douglas fir (*Pseudotsuga menziesii*) another western North American species, is capable of excellent growth on good sites, with yields which can exceed those from Sitka spruce. However, planting is limited by site conditions as Douglas fir only succeeds on well drained, fertile soils at low altitudes. It is nowhere common but is widely planted in western Britain, particularly in south-west England, Wales and the lower slopes of some of the western Scottish glens.

European larch

European larch (*Larix decidua*) is native to the Alps and other mountain ranges of central Europe but has been an important forest tree in Britain for more than 150 years. It is best adapted to the drier parts of the country where it is grown in plantations, shelter belts and as an estate and farmland tree. Though once the most abundant larch available, for many years Japanese and hybrid larches have been more commonly planted and, in the future, European larch is expected to provide only a modest volume and decreasing proportion of the British timber supply.

Larches are light-demanding trees, growing quickly when young. In forest crops, the lower branches die early and, unless removed, give rise to the small encased knots which are a feature of much larch timber. Larches grow well at high altitudes but the trees commonly respond to prevailing winds by producing a swept lower stem which causes problems on conversion, with the wood tending to spring from the saw.

Japanese larch

Japanese larch (*Larix kaempferi*) is better adapted than European larch for growth in the higher rainfall areas of the west and north and on less fertile soils. Most larch plantings in western Britain in recent decades have been of Japanese or hybrid larch owing to their faster growth and greater resistance to larch dieback compared with European larch. Future timber supplies of larch will be mainly Japanese larch.

Hybrid larch

Hybrid larch (*Larix* × *eurolepis*), the cross between Japanese and European larch (sometimes called Dunkeld larch) outgrows either parent on all but the very best larch sites and its superiority becomes very marked under conditions that are marginal for growth of larch. However, because of limited seed supply, planting of hybrid larch has been at a low level and timber supply for some decades will be small in comparison with that of Japanese larch.

Minor species

The most important of the so-called 'minor species' – western hemlock, grand fir, noble fir and western red cedar – were the subject of an earlier Forestry Commission Bulletin (Aldhous and Low, 1974).

Production

Annual production of the nine major softwood species is expected to increase from current production of over four million cubic metres to more than seven million cubic metres by the year 2000.

Well over half of the total production will be of Sitka and Norway spruce (Figure 1). Forecasts of production are reviewed regularly to take account of changes in the growing stock and in felling practices.

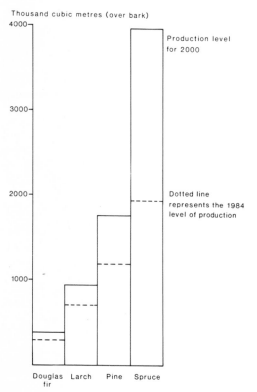

Thousand cubic metres (over bark)

Production level for 2000

Dotted line represents the 1984 level of production

Douglas fir Larch Pine Spruce

Figure 1. *Estimated roundwood production for the years 1984 and 2000.*

The Properties of British Softwoods

Most of the technical information on the properties of the important British softwoods has been derived from studies undertaken at the Princes Risborough Laboratory of the Building Research Establishment.

The properties of any timber show considerable variation depending on its seed origin, conditions of growth, age, etc. Over the past 30 years particular attention has been paid to the methods of sampling material for studies carried out at Princes Risborough for the Forestry Commission. Special care has been taken to obtain material for study which is representative of the timber likely to be available on the market. The joint programme has embraced all aspects of properties and utilisation, including a series of studies on the stress grading of Scots pine (Endersby and Hansford, 1971), Sitka spruce (Curry, 1973), Douglas fir (Fewell *et al.*, 1982), Corsican pine (Tory, 1978), and larch (Benham, 1986), an extensive investigation of 40–50 year-old Sitka spruce (Brazier *et al.*, 1976), and studies of the effects of spacing on the structural wood yields of Sitka spruce (Brazier *et al.*, 1985).

The basic properties, including density, strength, shrinkage, movement and other data for the commercially-important softwoods are given in Appendix I, and details of the kiln schedules referred to in Appendix III. A brief explanation of the terms used in classifying properties such as density, shrinkage and movement, natural durability, amenability to preservative treatment and working properties is given in Appendix IV.

Sitka spruce

The spruces are the only British softwoods of commercial importance which produce the almost uniformly pale-coloured wood described as whitewood. The timber is similar to whitewood from southern Scandinavia and parts of northern and central Europe.

Most British Sitka spruce is fairly fast grown, with timber which is light in weight and rather coarse in texture. Average density at 12 per cent moisture content is in the region of 390 kg m^{-3}. The light-coloured timber is of lustrous appearance and sometimes has a very pale pink or pinkish brown colour in the central core. When dry, there is no real distinction between heartwood and sapwood. The timber dries rapidly, but care is needed in drying if distortion (twist and cup) and degrade (splitting and loosening of knots) are to be minimised. There is, however, a wide variation in the extent of degrade on drying and also in the incidence of collapse. Kiln schedule J should prove satisfactory in most instances.

The strength properties of British Sitka spruce have been well characterised and provide a sound basis for its structural uses. Sitka spruce, like imported whitewood, is variable in strength and, although whitewood is given higher grade stresses in BS 5268 (Appendix II), Sitka spruce of similar strength can be selected by machine grading.

Sitka spruce is non-durable and both the sapwood and heartwood are difficult to treat with preservative solutions by pressure impregnation. However, the sawn timber can be treated successfully by the boron diffusion method, which has the advantage of being applicable to green timber as it comes off the saw. Boron treated timber is suitable for general building and construction work where protection against decay and insects is required, but it should not be used in ground contact or as unpainted components exposed to the weather since the preservative salt is subject to loss by leaching. Roundwood can be treated by sap displacement methods using copper/chrome/arsenic preservatives, but complete sapwood penetration is not always obtained.

The timber works easily with machine and hand tools and takes nails well. It does not finish as well as pine when machined and very sharp tools are needed to obtain a good finish with light weight wood or where there are broad bands of earlywood; such wood also tends to crumble when cut on the end-grain with a chisel. For most purposes for which British Sitka spruce is used, an acceptable finish can be obtained if cutters are maintained in a good condition.

Sitka spruce is used for a wide variety of applications including building, poles, mining timber, fencing, sheds and agricultural buildings. Its light weight, strength, nailability and non-tainting characteristics make it ideal for boxes, crates, pallets and other forms of packaging, particularly for fruit, foodstuffs and similar commodities. Because of its pale colour, good fibre characteristics and low resin and other extractive content, spruce is the preferred wood for making high quality mechanical pulp used in newsprint and other paper products.

Plate 2. *Timber testing – Bending strength of a beam being assessed at the Princes Risborough Laboratory.* (Copyright: Building Research Establishment)

Norway spruce

Timber properties are similar in many respects to those of Sitka spruce. The wood is typically of a cream colour throughout and has a natural lustre. Average density is similar to that of Sitka spruce at about 390 kg m^{-3} at 12 per cent moisture content.

The timber dries rapidly and well, with little tendency to split or check but, like Sitka spruce, with some risk of distortion. Kiln schedule K is advised, although more severe conditions can be tolerated.

The timber has marginally better strength and stiffness properties than Sitka spruce, but for design purposes both species are grouped together in BS 5268 : Part 2 : 1984 and given the same working stresses (Appendix II, Table A2).

It is non-durable and resistant to preservative treatment by pressure methods but the sawn, green timber can be readily treated by the boron diffusion method. An effective penetration of the sapwood can be obtained when roundwood is treated by sap displacement methods.

Norway spruce has very similar working properties to those of Sitka spruce but in general cuts more cleanly.

For most applications, the spruces are used interchangeably, but Norway spruce is the more widely used for power transmission poles since the sapwood is more easily treated than that of Sitka spruce.

Scots pine

Scots pine is the same species as the timber imported from northern Europe as redwood and is similar in character to much of the redwood originating from southern Scandinavia, Poland and areas close to the southern Baltic. The annual rings are clearly marked by the contrasting light-coloured earlywood and the dark latewood zones. The pale, reddish-brown, resinous heartwood of the timber, when dry, contrasts with the lighter-coloured sapwood, usually 50–100 mm wide. The average density of the timber at 12 per cent moisture content is about 510 kg m^{-3}.

The timber dries very rapidly and well, with very little degrade, but, because of its susceptibility to blue stain, logs should be converted and the sawn timber loaded into the drying kiln without delay. If this is not possible, blue staining of the sawn timber can be minimised by the application of an anti-stain treatment. Kiln schedule M is recommended (schedule F should be used if it is important to retain the original colour).

There is little difference in strength properties between Scots pine and imported redwood and, for structural purposes, they have the same grade stresses in BS 5268 : Part 2 : 1984.

The wood is non-durable and, although the heartwood is moderately resistant, it can be treated effectively with preservatives; the sapwood is particularly permeable to preservatives and full penetration can be readily achieved.

The timber works easily and cleanly in most hand and machine operations, but sharp cutting edges are required for the faster grown material with its wider bands of soft earlywood. Knots can also be troublesome when the timber is dry as they are liable to become loose and fall out during planing and sawing. Finishing and gluing are generally satisfactory, although some problems may arise with very resinous timber. Nailing is also satisfactory.

Scots pine is a versatile timber, suitable for a wide range of uses. It is used for many structural purposes, including some of the more demanding applications such as trussed rafters. It is also suitable for joinery and furniture.

Although not a durable timber, it is easily treated with preservatives and is ideally suited to applications where there is a high decay hazard, such as sleepers, fencing, posts and poles. It is the preferred species for telephone and power transmission poles. It is very suitable for the production of wood wool cement slabs because it shreds well and does not inhibit cement setting.

Corsican pine

British Corsican pine is similar to Scots pine in appearance. It has a larger proportion of sapwood and is typically of more vigorous growth, particularly in juvenile wood. This can give the timber a somewhat coarser texture, with larger knots and more pronounced knot whorls. Like Scots pine, it dries rapidly and well, though it is subject to stain if drying is delayed. Kiln schedule M is recommended but a milder schedule, G, should be used if resin exudation is to be avoided.

For most structural purposes, Corsican pine and Scots pine can be used interchangeably. For comparable grades, Corsican pine has the same grade stresses as Scots in BS 5268 : Part 2, except for somewhat lower figures for modulus of elasticity in bending; it is also less resistant to fracture on impact.

Working properties, including nailing and finishing, are very similar to those of Scots pine, but the knots of Corsican pine are slightly less hard and usually hold better during cutting operations. Shrinkage and movement values are slightly less than in Scots pine.

Durability and treatability are similar to those of Scots pine, but Corsican pine contains a larger amount of permeable sapwood.

Corsican pine is used for the same purposes as Scots pine, but is less commonly used for transmission poles.

Lodgepole pine

Lodgepole pine was not planted extensively in Britain until the 1950s and the information on the properties of the British timber has necessarily been derived from fewer samples of younger material, mainly grown on better than average sites.

The timber is a pale, straw colour, with heartwood which is difficult to distinguish from the comparatively narrow band of sapwood. It is soft, straight-grained and, having little contrast between the earlywood and latewood, is of fine, fairly uniform texture. Knots are generally small and tight. The average density at 12 per cent moisture content is in the region of 460 kg m^{-3}, which is intermediate in weight between Scots pine and spruce. Limited tests on British lodgepole pine indicate that it dries very rapidly and well, with no checking or splitting. The timber is normally free from distortion on drying, which is rated as mainly slight; resin exudation is fairly common. The rather severe kiln schedule M can be used, but schedule L is recommended.

Lodgepole pine has strength properties, based on small, defect-free pieces, between those of Scots pine and spruce; design stresses have yet to be derived and await the availability of adequate structural-sized material. The heartwood of lodgepole pine is non-durable and more resistant to preservative treatment than that of Scots pine.

The timber works easily with hand and machine tools and does not have an undue dulling effect on cutters, although resin exudation can sometimes be troublesome. It finishes cleanly, takes nails satisfactorily, gives fairly good results with paint and can be glued satisfactorily.

Because most lodgepole pine plantings are still young, sawnwood has yet to appear on the commercial market. Test material has shown British lodgepole pine timber to be similar to that of Scots pine but of somewhat lighter weight and with a finer texture. Sawnwood offers scope for use in building, for construction and some types of joinery, as well as for pallets, packaging, fencing, mining timber, etc., and small roundwood for chips and pulp.

Douglas fir

Douglas fir has a light reddish-brown heartwood, clearly distinguished from the comparatively narrow band of pale-coloured sapwood. The marked contrast between earlywood and latewood gives rise to a prominent growth ring figure, particularly evident on plain-sawn surfaces and rotary-cut veneers. The wood is generally straight-grained, but can sometimes display a tendency to wavy or spiral grain. British Douglas fir is less resinous than that grown in North America. The average density of the timber is about 530 kg m^{-3} at 12 per cent moisture content.

The timber dries rapidly and well, with little checking or distortion, but knots tend to split and loosen. When dried, it is particularly stable in use. Kiln schedule K gives satisfactory results.

It is a strong wood and, while not quite the equal of pine in bending strength, it is comparable in stiffness.

The timber is moderately-durable and, being resistant to preservative treatment, is more difficult to treat than pine. Incising the timber before treatment gives satisfactory results with items of larger cross-section, such as railway sleepers. Given an effective preservative treatment, Douglas fir will give excellent service in demanding exterior conditions.

The timber works readily with hand and machine tools, but not as easily as pine, and it has a greater dulling effect on cutters. Hard, loose knots are liable to damage cutting edges. A good finish can generally be obtained provided that sharp cutting edges are maintained, although there is a tendency for fast-grown material to splinter and break away at the tool exit where the cut is across the grain. In planing and moulding, dulled cutters tend to drag and compress the soft earlywood, which later expands and produces ridged surfaces. The timber takes nails and screws satisfactorily, but care is needed to avoid splitting. It stains effectively and gives good results with various finishing treatments when normal care is taken to prevent grain-raising. Material with a high resin content should be kiln-dried if it is to be varnished or painted for indoor purposes. The timber can be glued satisfactorily.

Douglas fir is available in large sections and long lengths and is primarily a construction wood. Selected timber is used for joinery purposes, where its stability in service and appearance are particularly advantageous.

Larch

There is little difference, technically, between the wood of the larches when a comparison is made between material of comparable age and grade. However, hybrid larch can grow vigorously under particularly favourable conditions, with a consequent reduction in wood density and in the characteristic

Plate 3. *Timber testing – Durability is an important characteristic of timber. This 'graveyard' of samples is used to find the average time to failure of heartwood in ground contact.* (Copyright: Building Research Establishment)

strength and decay resistance of larch.

The resinous, pale reddish-brown to brick red heartwood contrasts sharply with the narrow, light-coloured sapwood. Annual rings are clearly marked by the light earlywood and dark latewood zones. A characteristic feature of larch is the presence of many fairly small but often dead knots. Timber of old growth larch is among the heaviest of the British softwoods, averaging about 590 kg m^{-3} at 12 per cent moisture content but that from younger trees is somewhat lighter at about 510–540 kg m^{-3}.

The timber dries fairly rapidly but with some tendency to distort, split and check, and for knots to split and loosen. Kiln schedule H is recommended.

Larch is among the hardest and toughest of the British softwoods. In other strength properties it is similar to Scots pine and has the same working stresses in BS 5268. Differences between the timbers of the larch species are small and usually result from differences in age. More European larch timber comes from older trees and tends to be heavier, harder and stiffer, but such differences can be expected to diminish as older growth Japanese and hybrid larch becomes available.

Larch is the most durable of the British softwoods, but is resistant to preservative treatment; even the sapwood is moderately resistant. It has a tendency to spring off the saw on conversion, incurring some wastage. The dried timber saws and machines fairly easily and finishes cleanly in most wood-working operations, but the hard knots have a blunting effect on the cutting edges of tools. It takes stain, paint and varnish satisfactorily. Pre-boring is necessary to mini-mise splitting on nailing. Japanese larch is milder to work than European larch and, although the soft earlywood zones are liable to tear or crumble, a clean finish can normally be obtained if sharp cutting edges are used.

Larch has traditionally been regarded primarily as an estate timber, used for a variety of purposes such as buildings, fencing, gates, posts, etc., and is widely used for panel, interwoven and lap fencing in residential areas. It is the preferred softwood timber for boat-building. An appreciable volume of larch will become available for construction purposes over the next 20 years, although its hardness and tendency to

Plate 4. *Boat building – Boatskin larch is required by this industry where skippers still specify wooden hulls for their fishing boats such as this one at Jones Buckie Shipyard Ltd.* (E5919)

distort and to split on nailing could present problems. Greater use may be found in applications requiring larger sizes, where the inherent strength and durability of larch give it a distinct advantage.

Sawn Timber

Production

There are over 400 sawmills in Britain, ranging from small estate mills sawing less than 1000 m³ a year, to large modern mills with an input in excess of 50 000 m³ a year. However, about 65 per cent of the total output of sawn softwood is produced by about 18 per cent of the sawmills.

Softwood sawlog production is expected to increase from its present level of about 2.4 million m³ to about 4.0 million m³ per annum by the end of the century. Expansion of the production capacity of the larger existing sawmills is expected to cope with this increased volume.

Table 1. Estimated average annual sawn timber production (thousand cubic metres)

	1987/91	1992/96	1997/2001	2002/2006
Sitka spruce	415	480	665	880
Norway spruce	195	225	300	395
Scots pine	220	275	290	300
Corsican pine	70	80	100	100
Lodgepole pine	–	5	25	25
European larch	55	50	45	55
Japanese larch	130	175	210	265
Hybrid larch	15	25	30	40
Douglas fir	75	80	110	145
TOTAL	1175	1395	1775	2205

Source: Forestry Commission

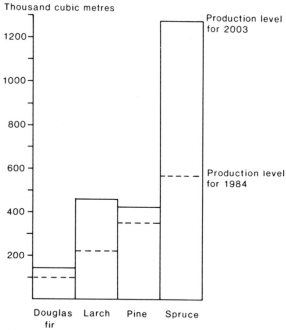

Thousand cubic metres

Production level for 2003

Production level for 1984

Douglas fir / Larch / Pine / Spruce

Figure 2. Estimated sawn timber production for the years 1984 and 2003.

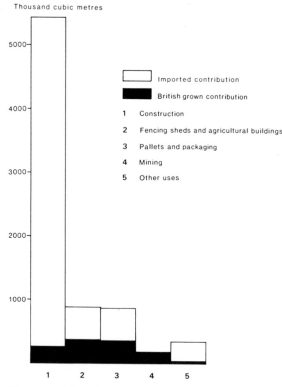

Thousand cubic metres

☐ Imported contribution

■ British grown contribution

1 Construction
2 Fencing sheds and agricultural buildings
3 Pallets and packaging
4 Mining
5 Other uses

Figure 3. Estimated annual sawn timber consumption, mid-1980s (Elliott, G.K., 1985).

The pattern of estimated production of sawn softwood is illustrated in Table 1 and Figure 2, but may be affected by fluctuations in total roundwood production and changes in conversion practice. The anticipated production is based on the following assumptions:

1. all roundwood with a diameter over bark of 18 cm and more, plus half of the volume between 14 and 18 cm over bark, is sold as sawlogs;
2. sawn out-turn is 51 per cent of over bark volume in relation to current production forecasts;
3. the figures for hybrid larch and Japanese larch are calculated pro rata according to the planted area (approximately 10 per cent hybrid, 90 per cent Japanese) because the production forecasts do not distinguish between the two species.

The significant aspects of the estimated production totals are that output will nearly double by the end of the century, and at that time over half of the output will be of Sitka and Norway spruce.

Current uses and prospects

In 1985, British sawn softwood accounted for an estimated 18 per cent of the total consumption (Forestry Commission, 1986). The construction industry is by far the largest market sector for sawn softwood, taking about 70 per cent of the total supply, but only 5 per cent of this is British-grown. Consumption by the main market sectors is illustrated in Figure 3 (Elliott,

1985). The 'other uses' category includes a variety of miscellaneous applications such as permanent way construction, vehicle and rolling stock manufacture and repair, boat building, furniture manufacture and joinery.

There is clearly scope for a significant increase in market penetration by British softwoods into the very large construction sector. The introduction of the strength class system of grading (BS 5268 : Part 2 : 1984) and the increasing attention being paid to good presentation of the sawn product, have already assisted in the wider use of British softwoods, particularly Sitka spruce, in construction. Indeed, the construction industry offers the best prospect of utilising significant quantities of the additional sawn timber production expected from British forests over the next 20 years.

Activity in the fencing, sheds and agricultural buildings sector is dependent upon levels of agricultural investment, consumer spending and road construction. In a much smaller market, and one in which British timber already has more than a 40 per cent share, there may be less scope for further penetration than into construction uses.

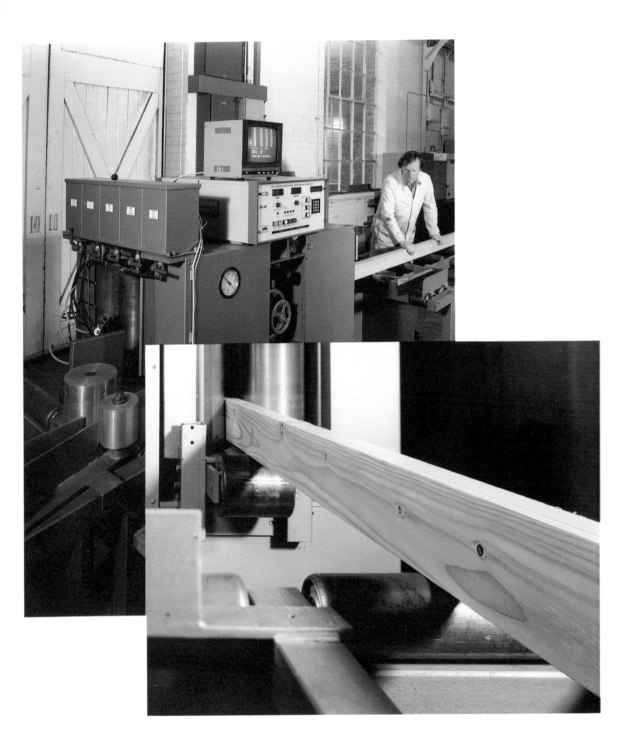

***Plate* 5.** *Grading – Machine Stress Grading is an objective method of characterising the strength of timber intended for structural use.* Inset *Detail of a larch batten passing through a stress grading machine.* (Copyright: Building Research Establishment)

The demand for crates, cases and cabledrums is not expected to increase by much, but the pallet sub-sector could absorb a higher proportion of British timber, although this would need to displace Portuguese maritime pine and the lower grades of European redwood and whitewood.

The demand for mining timber has been declining for many years and is likely to continue its downward trend.

British softwoods are little used for furniture and joinery, but selected Scots pine and Douglas fir can meet the more demanding requirements of these markets. The DIY market offers a further outlet for the use of British softwoods for a range of applications in and about the house and garden.

Requirements for use and suitability of the timbers

Building and construction

REQUIREMENTS

Several timber properties are important for the use of timber in building and construction, including the ease with which it can be worked, its dimensional stability, and in certain cases its natural durability or treatability with preservatives. In structures, strength and stiffness are of primary concern, and timber for this use is marketed in terms of its strength characteristics.

Because timber is variable in strength, systems of stress grading and strength classification have evolved. BS 5268 : Part 2 defines a series of five strength classes for softwoods but most commercial grades of structural softwoods fall into strength classes SC3 and SC4 (see Table A1 in Appendix II). Before any timber can be allocated to a strength class it must be graded either visually or by machine (BS 4978). Grading by machine is more efficient and provides higher yields, and for many British softwoods it is the only means of obtaining commercially viable yields of structural timber.

If British softwoods are to be marketed for structural use it is important that they are stress graded and that their position in the classification system is clearly understood. These aspects are explained more fully in Appendix II. It is also important that the expectations of the market are met in terms of moisture content, and dimensional tolerances, which are defined in British Standards (BS 5268; BS 4471 : 1987).

The requirements for construction are covered by a British Standard BS 5268 *Structural use of timber*. The seven parts to this standard are:

Part 1 Limit state design, materials and workmanship

Part 2 Code of practice for permissible stress design materials and workmanship

Part 3 Code of practice for trussed rafter roofs

Part 4 Fire resistance of timber structures

Part 5 Preservative treatments for constructional timber

Part 6 Code of practice for timber frame wall design

Part 7 Recommendations for the calculation basis for span tables.

These codes are primarily of interest to the designer but they carry implications for the supplier inasmuch as they permit or restrict the use of certain grades of timber.

Two other standards of importance to suppliers are BS 4978 which specifies the methods for visual and machine grading of structural timber and BS 4471 : 1987 which defines basic sizes for sawn softwoods and permitted deviations.

SUITABILITY OF THE TIMBERS

BS 5268 : Part 2 provides an objective basis for utilising timber structurally and all the British softwoods of commercial significance are included. Some timbers are intrinsically stronger than others and this is reflected in the strength class table. Thus the pines are stronger than the spruces and within a strength class (SC3) it is necessary to select a higher grade of Sitka spruce (M75) than Scots pine (GS) to meet the requirements of the class. Where a strength requirement cannot be met by a weaker timber it is, of course, always open to the designer to use a larger section size in compensation.

BS 5268 : Part 3 provides a Code of Practice for the fabrication and use of trussed rafters. Span tables are provided for Scots pine and Corsican pine graded to M75/SC5, SS/SC4, or M50. The timber must be dried to 22 per cent moisture content and there are close tolerances to be met on distortion as well as wane and fissures. M75 Sitka spruce can also be used provided it can meet the demanding quality requirements for these engineered products. Virtually all British softwoods can be used for general framing and carcassing purposes, provided they are appropriately sawn and graded. Where enhanced durability is required (BS 5268 : Part 5) it should be noted that Douglas fir and the larches are rated as moderately durable, and other species can be upgraded by preservative treatment. The boron diffusion process provides a method for treating green sawn timber; this is particularly useful for sawn spruce which is difficult to penetrate once it is dry. For the working characteristics of British softwoods see under 'The properties of British softwoods' pp. 8–13.

Plate 6. *Sawn softwood – A view over the sawmill yard of Western Softwoods Limited at Newbridge showing timber packs ready for despatch. (37644). Inset The SC3 grade mark on tested timber in the yard. (37647)*

***Plate* 7.** *Construction – British Douglas fir used as beams in the construction of a visitor centre for the Northumbrian Water Authority by A.G. Stuart Ltd.* (E8570)

With the forecast increase in production, Sitka spruce can be expected to displace a substantial proportion of imported whitewood currently used for general carcassing and framing. In doing this it will be competing with imported softwoods which fall into SC3 and SC4 and occasionally SC5. At present most Sitka spruce is able to provide acceptable yields of SC3, and some better quality material satisfies M75 and even SC4. There is good evidence that the yield of structural timber is directly related to the grade of sawlog. However, there is no room for any decrease in quality (there is little demand for SC2) and the emphasis should be on growing Sitka spruce with strength properties similar to those of imported whitewood.

Table 2. Strength classes, structural applications and suitable timbers

Strength class	Structural applications	Suitable timbers
SC5	Trussed rafters	Pine
SC4	Trussed rafters; general carcassing	Pine, larch* (see footnote)
SC3	General carcassing	Pine, spruce, larch, Douglas fir
SC2	General framing (less demanding applications)	Pine, spruce, larch, Douglas fir

*Larch cannot be used for trussed rafters at present because no Agrément Certificate for punched metal plate fasteners for use with larch has been issued.

Plate 8. *Construction – Machine graded Sitka spruce being used on a building site as joists in a suspended floor.* (37933)

Plate 9. *Fencing – Softwoods require treatment with preservatives to give a service life of 20 or 40 years. The post and rail fence round this paddock has been treated to the specification for 40 years.* (Copyright: Building Research Establishment)

Fencing

REQUIREMENTS

Requirements for the three main types of wooden fencing are covered by BS 1722 : Part 6 (pallisade); Part 7 (post and rail) and Part 11 (woven wood and lap boarded panel) (1986). Motorway fencing is covered by the Department of Transport's *Specification for highway works* (Department of Transport, 1986). The general requirements are the same as those detailed in BS 1722 : Part 7, relating to characteristics and defects such as size and distribution of knots, slope of grain, splits, checks, insect attack and rot, etc. However, the requirement for a 40-year service life for motorway fencing demands a high standard of preservative treatment according to BS 5589 : 1978.

For posts and other elements likely to remain in contact with the ground, a naturally durable timber is required, or one that if less durable, can be effectively treated. For some species, the sapwood is particularly easy to treat, and in such cases, a large proportion of effectively treated sapwood is advantageous.

The requirements for field gates and posts, set out in BS 3470 : 1975, are similar in many respects to those for fencing, but a generally higher quality of timber is required for gates.

SUITABILITY OF THE TIMBERS

Because of its strength and durability, larch is the popular softwood timber for all types of fencing, gates and similar farm and estate uses. Larch and Scots pine,

19

Plate 10. *Fencing – Motorway fencing for the Department of Transport has been a major market for British softwoods in the past 25 years.* (37736).

suitably treated, are commonly used for posts and all of the softwoods are used for fencing rails, again with appropriate preservative treatment. Spruce is not included in the Department of Transport specification for motorway fencing posts, which require a 40-year service life, but can be used in applications requiring up to a 20-year life, subject to correct treatment. Though the pines do not have a high natural durability, they are easy to treat with preservatives and the treated timber is very suitable for situations where there is a high decay hazard. Larch, on the other hand, has a higher natural durability than pine but is much more resistant to preservative treatment. BS 3470 : 1975 lists European larch, Scots pine and Douglas fir as being suitable for field gates and posts.

Farm buildings

REQUIREMENTS

The requirements for the various types and classes of farm buildings are set out in the relevant sections of BS 5502 : 1978–86. Farm buildings are classified by design life for structural components of 50, 20, 10 and 2

years, covering classes 1, 2, 3 and 4 respectively.

Normal design considerations apply to all structural elements, but in many situations there is the additional factor of decay hazard. The timbers most likely to decay are those in contact with soil or manure, or those in buildings with a high condensation risk, such as animal houses, glasshouses, drying or crop storage buildings. Timber used in these situations must be naturally durable or preservative-treated to the correct specification for the required service life/building class. The various types of preservative treatment and the treatments required for specified conditions of use are detailed in BS 5502 : Section 3.2.

SUITABILITY OF THE TIMBERS

The pines and spruces, suitably treated, are used extensively for vertical space boarding and for purlins. The pines are easily treated with preservatives and can be used, suitably treated, in high decay risk situations. The appropriate treatment schedules for spruce render them suitable for up to 20-year ground contact exposure or similar high risk situations. BS 5502 : Section 3.2 lists larch and Douglas fir as being suitable for

Plate 11. *Farm buildings – A pole barn at Bridgitt's Experimental Husbandry Farm near Winchester, designed by ADAS. (37729). The building was constructed using large treated round timber uprights and a novel system with short small diameter roundwood struts in the roof (Inset)*

constructional purposes, without preservative treatment, for a 20-year design life, in situations exposed to the weather, but not in ground contact or where the wood is liable to remain damp for long periods. However, it is important to note that this applies to the heartwood only – appropriate preservative treatment is required if sapwood is present.

Pallets

REQUIREMENTS

Requirements for pallets for materials handling for through transit are covered in BS 2629 : Part 1 : 1967; Part 2 : 1970; Part 3 : 1978. The original materials specification (Part 1 : 1967) was amended in 1978 and simply states that "all materials shall be of good commercial quality suitable for the method of construction so as to ensure that the pallet meets the requirements of Part 3 of this standard". Part 3 sets out performance requirements and methods of test for complete pallets, not individual components.

There are no directly applicable methods of assessing pallet board strength by visual or mechanical stress grading because of the comparatively thin board sections employed. If pallet failures are to be avoided, it is essential to ensure by quality control that those characteristics which diminish board strength are not excessive e.g. the size, occurrence and position of knots and knot whorls; slope of grain; amount of wane; decay or insect attack.

Ease of nailing, resistance to splitting on nailing and nail-holding ability are important aspects, though each is affected by the method of nailing and type of nail used.

Nail-popping and the loosening of joints can be minimised by fabrication at the appropriate moisture

Plate 12. *Pallets – A major use of small dimension British softwood logs is timber for the pallet and packaging market.* (E7993)

content and by using timbers with moderate or small 'movement' characteristics (see Appendix IV).

Deckboards may require the use of species with a high resistance to indentation if they are to be used for items such as kegs and drums which impose concentrated loads on the board surfaces. For some more extreme conditions of service, the leading edge boards require a high resistance to impact. Baseboards normally require higher strength than the deckboards because there are fewer boards to share the load and chamfering the top edges reduces the board strength. Indentation resistance may also be important. Stringers are subject to high nailing density along the board and good nailing properties are essential. They also require higher strength than deckboards. Blocks and bearers should have adequate strength combined with good nail-holding and high resistance to splitting on nailing during assembly and during drying in use.

Further guidance on the requirements for pallet boards is given in TRADA Wood Information Sheet 5/2 (Timber Research and Development Association, 1984).

SUITABILITY OF THE TIMBERS

All nine species can be used in pallet manufacture, provided that due regard is given to the occurrence of the strength-reducing characteristics already mentioned.

Because of their strength, nailing and nail-holding properties, the pines are used extensively in pallet manufacture, including blocks and bearers. The spruces are the lightest in weight of the timbers used for pallets, but their good strength-to-weight ratio makes them a preferred species where a light weight pallet is required. They are less resinous than the pines, and the typically small, tight knots do not normally affect pallet performance.

Larch is the strongest of the British softwoods used in pallet manufacture, but it does not take nails and other fastenings as easily as the pines and spruces. It is often used for leading edge boards because of its good impact resistance.

The properties of the timbers in relation to their suitability for pallet manufacture are summarised in Table 3.

Table 3. Properties of timbers significant for pallet boards

Species	Average density at 20% moisture content (kg m^{-3})	Ease of nailing	Resistance to splitting on nailing	Nail holding	Shrinkage on drying from green to 20% moisture content		Straightness on drying	Movement in service
					Radial %	Tangential %		
Scots pine	530	Fairly easy	Good	Good	1.6	2.5	Good	Medium
Corsican pine	530	Fairly easy	Good	Good	1.6	3.0	Good	Small
Lodgepole pine	480	Fairly easy	Satisfactory	Satisfactory	1.5	2.2	Good	Small
Douglas fir	550	Moderately difficult	Poor	Good	1.4	2.2	Good	Small
European larch	560	Moderately difficult	Poor	Good	1.6	2.5	Variable	Small
Japanese larch	530	Moderately difficult	Poor	Good	1.1	1.7	Variable	Small
Hybrid larch	500	Moderately difficult	Poor	Good	–	–	–	Small
Sitka spruce	410	Easy	Good	Satisfactory	1.6	2.8	Variable	Small
Norway spruce	410	Easy	Good	Satisfactory	1.1	2.2	Variable	Medium

Plate 13. Sawn mining timber – a traditional use for British softwoods. All timber supplied to British Coal is subject to a Quality Assurance Scheme, BS 5750. (E8303)

Sawn mining timber

REQUIREMENTS

Specifications covering the requirements for the various categories of sawn mining timber are issued by British Coal (formerly the National Coal Board). To satisfy quality assurance requirements, British Coal insist that all suppliers, including sub-contractors, seek registration under BS 5750 : Part 2 (1987), the UK national standard for quality systems, for most of their mining timber categories.

Although a proportion of hardwoods is used for certain categories of sawn mining timber, the information given below refers to softwood only.

The main categories of sawn softwood mining timber are:

1. *Chock timbers* are covered by British Coal Specification No 666/1985 (superseding No 666/1981) (British Coal, 1985a). Chock timbers currently account for over half of all sawn mining timber by volume and value, 1986 consumption being about 228 000 m^3, half of which was softwood.
2. *Boards for self advancing supports (SAS boards).* A British Coal Specification is in preparation. Current demand is about 40 000 m^3 per annum and rising.
3. *Sleepers* are covered by British Coal Specification No 537/1973 (British Coal, 1986). British Coal policy is to use softwoods only, and consumption in 1986 was around 12 000 m^3.

4. *Miscellaneous uses*, including crowns/baulks, pillarwood, coverboards (British Coal, 1984) and lids/blocks are covered by various British Coal Specifications now in preparation. These miscellaneous uses accounted for about 55 000 m³ in 1986.

The basic requirements for uses other than chocks are that they shall be of sound quality, free from loose and/or unsound knots, decay, insect attack and excessive shakes (splits) likely to affect their strength in use. This general specification is varied somewhat for chock timbers in that splits and shakes shall be limited to both 50 per cent of the length and 50 per cent of the thickness, and loose knots, knotholes and knots containing decay shall not exceed 25 mm in diameter and 25 mm in depth; sound knots are acceptable.

SUITABILITY OF THE TIMBERS

The pines, spruces, larches and Douglas fir are suitable for all of the sawn mining timber applications. British Coal prefer to use timber grown in Britain for all of their requirements.

Joinery

Although joinery accounts for only a small proportion of British softwoods (it is, in fact, included in the 'other uses' category in Figure 3), it is nevertheless considered worthy of mention.

Timbers for joinery must finish well, be stable in use and, if used out of doors, be durable or accept preservative treatments. The requirements for timber in joinery are set out in BS 1186 : Part 1 : 1986 which specifies limits for knot size and frequency, slope of grain and rates of growth. These requirements are difficult to meet with short rotation, plantation-grown timber and the small proportion of British softwood suitable for joinery is obtained from older and larger trees, mainly of Scots pine and Douglas fir.

Typical uses for sawn British softwoods

The typical uses for sawn British softwoods are summarised in Table 4.

Table 4. Typical uses for sawn British softwoods

Species	Construction	Pallets	Farm and estate use	Sawn mining timber	Joinery
Scots pine	Trussed rafters, general carcassing and framing	Deckboards, baseboards, stringers, blocks and bearers	Fencing, gates, posts, space boarding, purlins, framing	Chocks, coverboards sleepers and miscellaneous uses	Selected timber
Corsican pine	As above	As above	As above	As above	As above
Lodgepole pine	General carcassing and framing	As above	As above	As above	As above
Sitka and Norway spruce	General carcassing and framing	Deckboards, baseboards and stringers for lightweight pallets	Fencing rails, gates, space boarding, purlins	As above	
Douglas fir	General carcassing and framing	Deckboards, baseboards, stringers, blocks and bearers (nailing can be a problem)	Fencing, gates, boarding, purlins, framing	As above	Selected timber
Larch	General carcassing and framing	Deckboards, baseboards, stringers, blocks, bearers; often used for leading edge boards (nailing can be a problem)	Fencing, gates, posts, boarding and most farm and estate uses	As above	

Small Roundwood and Poles

Production and use

The term 'small roundwood' is normally used to indicate those stems, or parts of the stem, which are below the minimum diameter limit acceptable for sawlogs (between 14 and 18 cm) but usually greater than 6 cm in diameter. It is the product of thinnings, small final crop trees and the tops of larger trees. Although the character of the wood differs between these sources, they are rarely separated commercially.

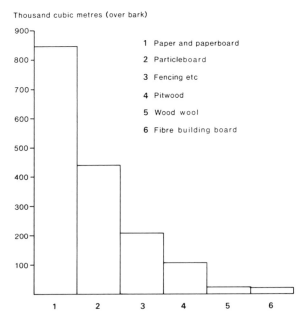

Thousand cubic metres (over bark)

1 Paper and paperboard
2 Particleboard
3 Fencing etc
4 Pitwood
5 Wood wool
6 Fibre building board

Figure 4. Deliveries of coniferous small roundwood, 1985.

Coniferous small roundwood is used for a variety of purposes and deliveries to the principal markets in 1985 are indicated in Figure 4. In addition to its use in the major wood processing industries, producing particleboard, fibre building board and paper and paperboard products, it is used extensively for farm and estate work and is stocked by many local merchants for garden use, sometimes as poles but also as round, or half-round posts. Small roundwood is mainly sapwood and, when used in ground contact or in other conditions favouring decay, an effective preservative treatment is essential to ensure an extended life. Pitwood continues to provide an important market for small roundwood.

The market for small roundwood has undergone significant changes in recent years. A major restructuring of the wood processing industry has seen the closure of obsolete plant and large-scale re-investment in new processing technology and products. The locations of the major British users of small roundwood, the panel products and pulp board mills, are shown in Figure 5.

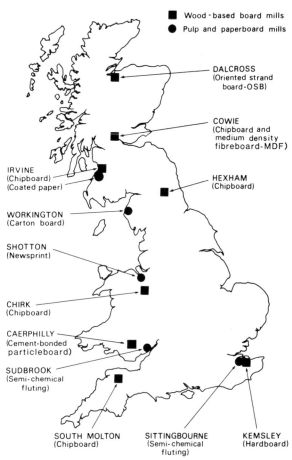

Wood-based board mills
Pulp and paperboard mills

DALCROSS
(Oriented strand board-OSB)

COWIE
(Chipboard and medium density fibreboard-MDF)

IRVINE
(Chipboard)
(Coated paper)

HEXHAM
(Chipboard)

WORKINGTON
(Carton board)

SHOTTON
(Newsprint)

CHIRK
(Chipboard)

CAERPHILLY
(Cement-bonded particleboard)

SUDBROOK
(Semi-chemical fluting)

SOUTH MOLTON
(Chipboard)

SITTINGBOURNE
(Semi-chemical fluting)

KEMSLEY
(Hardboard)

Figure 5. Locations of the major wood processing plants.

An important roundwood use, though from stems of sawlog size, is for telephone and electricity transmission poles. The annual requirement is for about 130 000 poles (approximately 36 000 m³), of which around 50 000 (about 10 000 m³) mainly of Scots and Corsican pine are currently supplied from British forests. Though criteria of length and stem form are demanding, poles offer a good return for the grower. The requirements for use and suitability of British timbers for transmission poles, are dealt with in more detail on pp. 31–32.

Plate 14. *Pulp and paper – A view over the log yard at Shotton Paper Company Limited in North Wales.* Inset *A roll of finished paper leaving the end of the production line.* (Copyright: Shotton Paper Co. Ltd.)

Requirements for use and suitability of the timbers

Paper and paperboard

REQUIREMENTS

The production of pulp for paper and paperboard products is currently the largest single market for small roundwood. The purpose of pulping is to separate the wood fibres so that they can be re-formed into sheets of paper, paperboard, carton board, etc., and the preferred species are those with pale-coloured wood, long, strong fibres and low resin and other extractive content. The processes used to achieve fibre separation are:

1. Mechanical, in which a fibre pulp is obtained using a grindstone (groundwood pulp) or by the defibration of wood chips by means of rotating grooved discs (thermo-mechanical pulp). Yields are high, at around 95 per cent, but the resultant paper is weak and is normally used for newsprint. Mechanical pulp is usually blended with a proportion of stronger pulp when a better quality product is required.

2. Chemical, in which a fibre pulp is obtained by cooking the wood chips with chemical solutions to remove most of the lignin and hemicellulose. Though pulp yields are low, at around 50 per cent, the paper produced has a high strength.

3. Semi-chemical, in which softening of the chips is achieved by a mild chemical treatment followed by mechanical separation of the fibres in a disc refiner. The pulp is used for corrugated board, cardboard, etc. Yields from the semi-chemical processes are intermediate between those obtained from the mechanical and chemical processes.

A more recent development, chemi-thermo-mechanical pulping (CTMP) gives high yields (91–95 per cent) of pulp suitable for a range of products such as tissues, paperboard, newsprint and printing paper. The process is not used at present in any of the British pulp mills.

SUITABILITY OF THE TIMBERS

The fibre characteristics of all British softwoods make them suitable for pulp production. Because of their chemical composition, some species are more amenable to pulping than others, but by judicious choice of pulping conditions and after-treatment (beating, bleaching, etc.) any of them may be processed successfully.

Because of its pale colour, good fibre characteristics and low resin and other extractive content, spruce is the preferred wood for high quality groundwood pulp used in newsprint and other paper products. A wider range of species is used for processes involving chemical treatment, including semi-chemical pulping, but the presence of extractives can cause problems in processing.

Particleboard — wood chipboard

The chipboard industry is the next largest consumer, after pulp, for small roundwood; it also uses a substantial quantity of sawmill residues and other waste wood.

REQUIREMENTS

Chipboard is specified in terms of property requirements for the manufactured product. These are given in BS 5669 : 1979 which places no restriction on the wood raw material used, which is normally negotiated between the manufacturer and supplier.

Wood properties are of relatively minor importance for their effect on board properties, which are determined more by aspects of manufacture, such as particle size and distribution and adhesive type. Wood density is of some relevance for its effect on cutters during the chipping process, and wood colour perhaps more so since, in many applications, a pale-coloured board is preferred and the amount of dark-coloured heartwood in the furnish must be restricted; bark, too,

Plate 15. Particleboard – Logs entering the Egger's factory at Hexham, Northumberland. (Copyright: Egger (UK) Ltd.)

27

Plate 16. *Newly manufactured wood chipboard being examined for defects in the Egger's factory.* (Copyright: Egger (UK) Ltd.)

detracts from the appearance of boards but this is usually removed from roundwood by the board manufacturer. The presence of extractives in the wood can adversely affect the setting of adhesives, leading to a reduction in internal bond strength.

SUITABILITY OF THE TIMBERS

All nine species are used to some extent in the production of wood chipboard, but manufacturers prefer the spruces and pines. Smaller quantities of larch and Douglas fir are used because of their darker colour and the presence of extractives. For this reason they are mixed with other species. In many cases, price and the availability of regular supplies are the main factors influencing the choice of raw material.

Oriented strand board (OSB)

A recent development in wood-based board materials is oriented strand board (OSB). This is similar in some respects to waferboard, which is manufactured from large flakes of wood bonded together with a phenolic resin. The flakes used in OSB, though as long as those used in waferboard, are much narrower; this gives improved efficiency in resin distribution, and allows some measure of alignment of the flakes in OSB so that layers within the board have a grain alignment somewhat like that of adjacent veneers in plywood. Currently in Britain, OSB is only made with pine.

Fibre building board (including medium density fibreboard – MDF)

REQUIREMENTS

Fibre building board is produced by the adhesion of fibres and fibre bundles to form sheets. This is achieved by one of two processes depending upon the board type required:

1. a wet process, in which a pulp slurry, after draining, is pressed using heat; the adhesive properties of the lignin bind the fibres to form a board,
2. a dry process, in which a dry fibre mix has resin added to bind the fibres under heat and pressure; this type of board is known as medium density fibreboard, or MDF.

The need to reduce the wood to fibre bundles requires a high power consumption, favouring the use of the lighter-weight woods. The presence of resin in the wood can be troublesome for its effects on the

28

Plate 17. Wood-based panel products –

Wood chipboard

Hardboard (a fibre building board)   *Oriented strand board (OSB)*

Cement bonded particleboard   *Medium density fibreboard (MDF)*

Wood wool cement slab

(Copyright: Building Research Establishment)

cutting and beating operations. Although the presence of bark is tolerated in some measure, it affects the manufacturing process and can also detract from the appearance and quality of the board. Several types of fibre building board are produced, differing primarily in the density achieved in pressing. BS 1142 : Parts 1–3 : 1971–72 provides a specification for the properties and methods of test for manufactured boards. The quality of the wood raw material is normally agreed between supplier and manufacturer, as in the case of chipboard.

SUITABILITY OF THE TIMBERS

All the softwood species are used to some extent in the manufacture of fibre building boards but the spruces are preferred to the more resinous species.

Pitwood

British Coal Specification 695 : 1985 (British Coal, 1985b) sets out the requirements for British softwood pitprops and split pitprops. These relate to dimensions, straightness, knots, peeling, butt flare and

Plate 18. *Poles – Preparation of poles at forest road side. These may be used to carry overhead lines for either electricity or telecommunications.* (E7759). *Inset These poles are ready for use after creosote pressure treatment at the Grange Court plant of Calders and Grandidge Ltd. Some Electricity Boards have accepted CCA treated spruce poles.* (37658)

seasoning. Any species of coniferous timber can be used. British Coal prefer to use British timber.

Wood wool

REQUIREMENTS

British Standard 2548 : 1986 specifies requirements for wood wool for general packaging purposes and BS 1105 : 1981 and BS 3809 : 1971 refer to manufactured items using wood wool. The most important properties for processing are light weight, straight grain and freedom from knots.

Tough but soft and resilient fibre is the main consideration for packing items such as machinery, hardware, glass, china, etc. For foodstuffs, fruit, upholstery, etc., freedom from odour, gums and resins is essential to avoid tainting the packaged items.

When used for wood woolcement slab manufacture, high extractive contents (sugar, hemicelluloses and some phenolic compounds) can retard or impair the setting of the cement. Decayed wood must not be used but the presence of blue stain does not have a marked effect on the cement-setting process.

SUITABILITY OF THE TIMBERS

The spruces and pines are the preferred species for wood wool products, the spruces because of their pale colour and low extractive content, the pines because they shred very easily. Colour and extractive content make the use of Douglas fir and larch less acceptable.

Transmission poles

REQUIREMENTS

The dimensions and quality requirements for softwood poles for overhead power and telecommunication lines are set out in BS 1990 : Part 1 : 1984.

Table 5. Relative strengths of pole species

Scots pine (redwood)	1.00
Corsican pine	1.00
Douglas fir	1.17
European larch	1.32
Japanese larch	1.20
Hybrid larch	1.08
Lodgepole pine	0.83
Norway spruce	0.84
Sitka spruce	0.82

Source: BS 1990 : Part 1

These relative strengths are based on tests carried out on 6 m lengths and relate to green timber not pre-treated by spraying or ponding.

However, the acceptable species, size categories, methods of preservative treatment and method of purchase vary between the two main consumer industries, telecommunications and electricity supply.

Design data for strength are based on requirements for redwood (Scots pine), taken as a base, and modification factors are applied for poles of the same size of other species (Table 5).

Long life is a prerequisite for poles and preservative treatment is essential. It is expected that poles complying with BS 1990 will have a service life of 40 years. Accordingly, the standard stipulates that the pines, larches and Douglas fir shall be pressure impregnated with creosote to BS 913 : 1973 or with a copper/chrome/arsenic mixture to BS 4072 : 1987 to achieve full sapwood penetration and the required net retention of preservative. These requirements are, in fact, difficult to obtain with Douglas fir and larch. Spruces cannot be adequately treated with preservatives in the normal way by the procedures described in BS 913 and BS 4072, but BS 1990 allows a modification of one of the procedures described in these standards, provided that the requirements for sapwood penetration and preservative retention can be met.

The sap displacement method of preservation, whereby the fluid contained in the sapwood of freshly-felled poles is displaced by water-borne preservatives, is used successfully for the preservation of spruce poles, particularly Norway spruce. Since this method is applicable only to poles in the green condition, rapid extraction and treatment are essential.

SUITABILITY OF THE TIMBERS

All nine species covered by this report are listed in BS 1990 : Part 1. Because of the demanding criteria for length and form, yields of suitable stems per hectare are generally rather low, except in the case of Corsican pine, which is often high, and Norway spruce, which is usually good.

Scots pine has long been the preferred timber for poles, and it has become the yardstick for assessing the performance of other species. It has good strength properties, and dry sapwood is readily treated with preservatives giving a more than adequate penetration and retention.

Corsican pine is used by both the telecommunications and electricity supply industries. It has strength and preservation properties comparable to those of Scots pine. It has a larger proportion of sapwood than other species, and while this provides the potential for deep penetration of preservative, it also means that more care is needed in drying to ensure that the pole is at the right moisture content for treatment.

Larch poles are usually stronger than those of Scots pine but are more difficult to treat with preservative.

Douglas fir poles are in regular use by British Telecom. They have similar strength properties to those of Scots pine, but they are more resistant to treatment with preservatives.

Norway spruce is used for power transmission poles but is currently not accepted by British Telecom. It has a lower bending strength than Scots pine, hence slightly stouter poles are required to give the same structural performance. It is extremely difficult to impregnate satisfactorily with preservatives by conventional techniques and sap displacement methods must be used.

Similar considerations apply to Sitka spruce except that, even when special techniques are used, it is difficult to achieve an adequate and uniform depth of penetration of preservative.

Plate 19. *Bark – An important residue from other processes, bark is used to fuel boilers. However, there is a lucrative horticultural market for composted bark for potting and with larger particles for mulching beds.* (37924)

Further information on the production of poles for electricity supply and telecommunications is given in Forestry Commission Forest Record 128 (Aaron and Oakley, 1985).

Conifer bark utilisation

The use of bark has shown a steady rise over the past 10 years. Bark products are currently used extensively for horticultural purposes – potting and rooting mixtures, orchid cultivation, bulb forcing, mulching and landscaping, plunge beds, mushroom culture, glasshouse crops and soil improvement.

Bark is of particular value when used as a litter for cattle because of its ability to absorb ammonia and other chemically active gases. Coarsely pulverised bark is also used for equestrian purposes such as riding school paddocks, showjumping arenas and racehorse training gallops.

Bark can be used as a fuel, but it has a higher ash content than most fuels and briquettes made from bark are not a smokeless fuel.

Bark absorbs oil more readily than it takes up water, and this has been put to good use in the control of oil pollution – long, tubular nets are filled with bark to form a boom which is deployed along the surface of the water to prevent the spread of oil.

It is technically feasible to use extractive-rich conifer bark as a source of tannin for the leather industry. However, the harvesting and preparation of tan bark is labour intensive in that it must be hand-peeled and air-dried, making it economically unattractive compared with imported vegetable tannin materials and synthetic materials.

A detailed account of conifer bark properties and uses is given in Forestry Commission Forest Record 110 (Aaron, 1982).

APPENDIX I
Summary of the Timber Properties

Species	Average density (kg m^{-3}) at 12% moisture content	Shrinkage: green to 12% moisture content		Movement	Recommended kiln schedule	Durability	Resistance to impregnation
		Radial %	Tangential %				
Scots pine	510	3	4.5	Medium	M (if darkening not desirable, schedule F)	Non-durable	Moderately resistant; sapwood permeable
Corsican pine	510	3	5.5	Small	M	Non-durable	Moderately resistant; sapwood permeable
Lodgepole pine	460	2.5	4	Small	L	Non-durable	Resistant
Sitka spruce	390	3	5	Small	J	Non-durable	Resistant
Norway spruce	390	2	4	Medium	K	Non-durable	Resistant
Douglas fir	530	2.5	4	Small	K	Moderately durable	Resistant
European larch	540	3	4.5	Small	H	Moderately durable	Resistant; (sapwood moderately resistant)
Japanese larch	510	2	3	Small	H	Moderately durable	Resistant; (sapwood moderately resistant)

APPENDIX II
Strength, Stress Grading and Strength Classes

While it is common knowledge that the timbers derived from different tree species (e.g. oak and pine) are different in character and strength, it is perhaps less well appreciated how much mechanical properties vary within a single species. This variability can be attributed to:

1. variations in the strength of the clear wood substance, and
2. variations due to the presence of knots, slope of grain and other features which reduce strength.

The designer could ignore the variability and use very low design stresses, but that would be wasteful. Instead, systems of stress grading have been developed which enable a parcel of timber to be divided into stronger and weaker members, based on visual or machine sorting.

The earliest methods for stress grading softwoods relied on visual inspection. Essentially *visual stress grading* sorts timber into groups according to rules based on the number and position of the visible strength reducing factors described in 2 above. BS 4978 : 1973 describes the system of visual grading currently in use and defines two grades – Special Structural (SS) and General Structural (GS). The variation in strength of clear wood cannot be taken into account in visual grading, although limitations are placed on extreme rates of growth.

It should be noted that SS and GS are grades, not stresses or strength values. They are simply qualifying labels and stresses can only be allocated when they are associated with a particular timber, e.g. SS Scots pine or SS Sitka spruce. Because Sitka spruce and Scots pine differ in strength, SS Sitka spruce has different stresses from SS Scots pine (Table A2).

In recent years *machine stress grading* has become more widely used since it has a number of advantages over visual grading. It is faster, less subjective and it produces higher grade yields. Most stress grading machines depend upon the relationship between the stiffness and strength of timber. The measurement of stiffness by the machine provides an estimate of the strength of the piece, which embraces all the variability covered by 1 and 2 above. For this reason it is more discerning than visual grading, and this accounts for the higher yields obtained.

In theory, timber graded by machine can be graded to any strength within its range. Two machine grades, MSS and MGS, are defined which have the same stresses as the visual grades SS and GS. In addition, two numbered machine grades, M50 and M75 are commonly specified, which relate to an older system of visual grading which has now been replaced. These numbered grades are still used where, for a particular timber, it is advantageous in terms of grade yield.

Machine grading is a valuable aid to marketing British softwoods for structural use, and for Sitka spruce it is regarded as essential if good yields of the better grades are to be obtained.

For each combination of grade and timber species (e.g. GS Scots pine), BS 5268 : Part 2 provides design stresses which can be used directly if required (Table A2). This presents the designer with a multiplicity of different stresses and to simplify his task a system of *strength classes* has been derived. This system enables the grade/species combinations with similar strength to be grouped together into classes to which common stresses are assigned. The strength class system, as it affects British softwoods, is illustrated in Table A1 and the appropriate stresses for strength classes SC1 (the lowest) to SC5 are given in Table A3.

Table A1 also includes some common imported softwoods so that it can be seen where British-grown softwoods can substitute for their imported equivalents.

Table A1. Strength classes: BS 4978 stress grade/species combinations grouped under the BS 5268 strength classes

Standard name	Strength classes				
	SC1	SC2	SC3	SC4	SC5
British					
Douglas fir		GS	M50/SS		M75
Larch			GS	SS	
Scots pine			GS/M50	SS	M75
Corsican pine		GS	M50	SS	M75
Norway spruce	GS	M50/SS	M75		
Sitka spruce	GS	M50/SS	M75		
Imported					
Redwood			GS/M50	SS	M75
Whitewood			GS/M50	SS	M75
Spruce-pine-fir (Canada)			GS/M50	SS/M75	

Key: Visual grades	Machine grades
SS – Special structural	M75 – Machine 75
GS – General structural	M50 – Machine 50

The strength class system gives the supplier some flexibility in meeting the designer's requirements. If a strength class is specified, and there are no other constraints on the material supplied, the designer's requirements can be met by supplying any timber of the grade specified for that class.

BS 5268 also allows timber to be machine graded directly to the strength class boundaries and this will often result in a better yield than that given by the grade/species combinations associated with that strength class. However, it is important to realise that the class boundary stresses are sometimes lower than those of the individual timber grades contained in the class and this has important consequences. Thus, if a designer specifies SC3 timber, then M75 or SC3 Sitka spruce could be supplied. However, if M75 Sitka spruce is specified, then SC3 cannot be substituted. Similarly, if GS redwood/whitewood (which has the same stresses as the SC3 boundary) is specified, approval must be obtained from the designer before anything other than GS redwood/whitewood is supplied, even though Sitka spruce graded to SC3 or M75 may appear acceptable. Not only may one particular strength property be critical, but there may be other considerations, such as amenability to preservative treatment.

Further information on the specification of structural timber is given in Building Research Establishment Digest No 287 (Building Research Establishment, 1984).

Table A2. Grade stresses for softwoods (graded to BS 4978 rules) for the dry exposure condition (extract from BS 5268 : Part 2 : 1984 Section two)

Standard name	Grade	Bending parallel to grain* $N\ mm^{-2}$	Tension parallel to grain* $N\ mm^{-2}$	Compression		Shear parallel to grain $N\ mm^{-2}$	Modulus of elasticity	
				Parallel to grain $N\ mm^{-2}$	Perpendicular to grain† $N\ mm^{-2}$		Mean $N\ mm^{-2}$	Minimum $N\ mm^{-2}$
Redwood/whitewood	SS/MSS	7.5	4.5	7.9	2.1	0.82	10500	7000
(imported) and	GS/MGS	5.3	3.2	6.8	1.8	0.82	9000	6000
Scots pine	M75	10.0	6.0	8.7	2.4	1.32	11000	7000
(British)	M50	6.6	4.0	7.3	2.1	0.82	9000	6000
Corsican pine	SS/MSS	7.5	4.5	7.9	2.1	0.82	9500	6500
(British)	GS/MGS	5.3	3.2	6.8	1.8	0.82	8000	5000
	M75	10.0	6.0	8.7	2.4	1.33	10500	7000
	M50	6.6	4.0	7.3	2.0	0.83	9000	5500
Sitka spruce and	SS/MSS	5.7	3.4	6.1	1.6	0.64	8000	5000
Norway spruce	GS/MGS	4.1	2.5	5.2	1.4	0.64	6500	4500
(British)	M75	6.6	4.0	6.4	1.8	1.02	9000	6000
	M50	4.5	2.7	5.5	1.6	0.64	7500	5000
Douglas fir	SS/MSS	6.2	3.7	6.6	2.4	0.88	11000	7000
(British)	GS/MGS	4.4	2.6	5.6	2.1	0.88	9500	6000
	M75	10.0	6.0	8.7	2.9	1.41	11000	7500
	M50	6.6	4.0	7.3	2.4	0.88	9500	6000
Larch	SS	7.5	4.5	7.9	2.1	0.82	10500	7000
(British)	GS	5.3	3.2	6.8	1.8	0.82	9000	6000

*Stresses applicable to timber 300 mm deep (or wide).

†When the specifications specifically prohibit wane at bearing areas, the SS grade compression perpendicular to the grain stress may be multiplied by 1.33 and used for all grades.

Table A3. Grade stresses and moduli of elasticity for softwood strength classes, for the dry exposure condition

Strength class	Bending parallel to grain	Tension parallel to grain	Compression parallel to grain	Compression perpendicular to grain*		Shear parallel to grain	Modulus of elasticity	
							Mean	Minimum
	$N\ mm^{-2}$	$N\ mm^{-2}$	$N\ mm^{-2}$	$N\ mm^{-2}$	$N\ mm^{-2}$	$N\ mm^{-2}$	$N\ mm^{-2}$	$N\ mm^{-2}$
SC1	2.8	2.2	3.5	2.1	1.2	0.46	6800	4500
SC2	4.1	2.5	5.3	2.1	1.6	0.66	8000	5000
SC3	5.3	3.2	6.8	2.2	1.7	0.67	8800	5800
SC4	7.5	4.5	7.9	2.4	1.9	0.71	9900	6600
SC5	10.0	6.0	8.7	2.8	2.4	1.00	10700	7100

*When the specification specifically prohibits wane at bearing areas, the higher values of compression perpendicular to the grain stress may be used, otherwise the lower values apply.

APPENDIX III
Kiln Schedules*

| | Timber moisture content % | Standard schedule conditions for timber thickness up to 38 mm | | | Modification to schedule for thicker timbers | |
| | | | | | 38–75 mm | over 75 mm |
		Temperature dry bulb °C	Temperature wet bulb °C	Relative humidity % (approx.)	Temperature wet bulb °C	Temperature wet bulb °C
SCHEDULE F						
Stage 1	Green	50	45	75	46	47
2	60	50	44	70	45	46
3	40	50	42	60	43	44
4	30	55	43.5	50	45	46
5	25	60	46	45	47.5	49
6	20	70	52.5	40	54.5	56.5
7	15	75	56.5	40	58.5	60.5
SCHEDULE G						
Stage 1	Green	50	47	85	48	49
2	60	50	46	80	47	48
3	40	55	51	80	52	53
4	30	60	54.5	75	55.5	57
5	25	70	62.5	70	64	65
6	20	75	62.5	55	64	65.5
7	15	80	61	40	63	65
SCHEDULE H						
Stage 1	Green	60	55.5	80	57	58
2	50	60	54.5	75	55.5	57
3	40	60	52	65	53	54.5
4	30	65	53.5	55	55	56.5
5	20	75	56.5	40	58.5	60.5
SCHEDULE J						
Stage 1	Green	60	53	70	54.5	55.5
2	50	60	50.5	60	52	53
3	40	60	47.5	50	49	50.5
4	30	65	48.5	40	50.5	52
5	20	75	52	30	54.5	56.5
SCHEDULE K						
Stage 1	Green	70	65	80	66.5	68
2	50	75	67	70	68.5	70
3	30	80	68.5	60	70.5	72
4	20	90	69	40	71.5	73.5
SCHEDULE L						
Stage 1	Green	80	72	70	73.5	75
2	40	90	69	40	71.5	73.5
SCHEDULE M						
Stage 1	Green	90	81	70	83	84.5
2	50	95	78	50	80	82

*Extracted from the *Timber drying manual* (Building Research Establishment, 1986).

APPENDIX IV
Terms Used in the Assessment of Timber Properties

Density

The density of a piece of wood depends on several factors. It varies with the amount of water it contains and for this reason it is important that when the density of a timber is stated, the moisture content at which the weight determination was made should be cited. The densities of the timbers when dry refer to a moisture content of 12 per cent. The density of a timber at any other moisture content within the range of, say, 5 per cent and 25 per cent can be estimated with fair accuracy by adding or subtracting 0.5 per cent of the given density for each 1 per cent moisture content above or below 12 per cent. In all species, a considerable variation in density is found to occur apart from that arising from differences in the amount of contained moisture and the average densities given are only approximate.

Shrinkage and movement

Shrinkage measurements in the tangential and radial directions are obtained when plain-sawn and quarter-sawn boards are dried from a green condition. Conventionally, shrinkage is expressed as the percentage reduction in the green dimension on drying to 12 per cent moisture content.

The term 'movement' refers to the dimensional change that takes place when seasoned timber is subjected to change in atmospheric conditions. Movement is determined by calculating the percentage change in width when test samples are conditioned first in air at 90 per cent relative humidity and then in air at 60 per cent relative humidity, at a constant temperature of 25°C.

It is important to note that shrinkage and movement are not directly related one to the other. For example, it is possible that a wood may shrink quite appreciably in drying from the green to 12 per cent moisture content, yet it may undergo comparatively small dimensional changes when subjected to a given range of atmospheric conditions in service.

Shrinkage values are useful for estimating roughly the dimensional allowances necessary in converting green material, though further allowances for possible losses arising from distortion and sawing accuracy must also be added.

Movement values give some indication of how the dried timber will tend to behave when subjected to atmospheric changes in service. A so-called 'stable' timber is one that exhibits comparatively small dimensional changes in passing from the 90 to the 60 per cent humidity conditions.

Natural durability

In this country the term durability generally refers to the resistance of a timber to fungal decay and it is used in this sense here. Durability is of importance only where a timber is liable to become damp, as, for example, where it is used out of doors. It is of no consequence where a timber can always be kept dry because, under these conditions, wood-destroying fungi are not active.

The durability of most timbers varies a great deal and even pieces cut from the same tree can show wide differences, so it is only possible to speak of durability in approximate terms. In this Bulletin it is described by means of five grades as follows:

Durability grade	Average life of 50 × 50 mm stakes in ground contact
Perishable	Less than 5 years
Non-durable	5–10 years
Moderately durable	10–15 years
Durable	15–25 years
Very durable	Over 25 years

The classification is primarily a relative one but, from the results of field tests being carried out at the Princes Risborough Laboratory, it has been possible to give in the table some quantitative meaning to each grade. Thus timbers of the very durable grade may be expected to have an average life of over 25 years when used in contact with the ground in this country. The life stated for each grade relates to material of 50 × 50 mm (2 × 2 in) section. Larger sizes will, of course, last longer, but the increase will depend on the kind of wood. In general, in larger sizes the durable woods will last much longer, but perishable ones only slightly longer, than the figures given. Timber used externally, but not in contact with the ground, will generally have a much longer life than that indicated in the above Table.

Except where otherwise stated, the durability given refers to heartwood; the sapwood of almost all timbers is either perishable or non-durable and may be susceptible to insect attack in service. It is essential to remember this when dealing with timbers which may sometimes contain a high proportion of sapwood.

Amenability to preservative treatment

The following terms are used to describe the extent to which a timber can be impregnated with preservatives.

Permeable

These timbers can be penetrated completely under pressure without difficulty, and can usually be heavily impregnated by the open tank process.

Moderately resistant

These timbers are fairly easy to treat, and it is usually possible to obtain a lateral penetration of the order of 6–19 mm in about 2–3 hours under pressure.

Resistant

These timbers are difficult to impregnate under pressure and require a long period of treatment. It is often very difficult to obtain more than about 3–6 mm lateral penetration. Incising is often used to obtain a better treatment.

Extremely resistant

These timbers absorb only a small amount of preservative even under long pressure treatments. They cannot be penetrated to an appreciable depth laterally and only to a very small extent longitudinally.

Working properties

The notes on the working properties of the timbers described in this Bulletin are based generally on the behaviour of normal kiln-dried material (moisture content from 10 to 14 per cent). It should be realised that, particularly with the more dense timbers, drier material has a greater resistance to cutting and an increased dulling effect on tools. The increased brittleness at the lower moisture content is, however, of assistance in planing timber having wavy or disturbed grain, as the chips break more easily and such tearing as occurs is less severe.

References to standard working conditions relate to operations carried out on ordinary commercial machines with cutter positions, speeds, etc., as provided by the manufacturer. The cutting angle of about 32° found on most planing machines is suitable for most softwoods. With a large number of softwoods, it is essential to maintain the tools in a sharp condition in order to cut the earlywood cleanly and to minimise the defect known as 'raised grain' on planed and moulded surfaces. The use of square cutterblocks and an efficient waste-removal system give improved results with those timbers that tend to chip-bruise in planing.

The remarks on nailing and screwing refer to the behaviour of the timber when normal care is taken to use nails of suitable size and to pre-bore correctly for screws. The notes on gluing similarly assume good conditions for application and setting.

References

British Standards

BS 913 : 1973 *Specification for wood preservation by means of pressure creosoting.*

BS 1105 : 1981 *Specification for wood wool cement slabs up to 125 mm thick.*

BS 1142 : Part 1 : 1971 *Specification for fibre building boards. Methods of test.* Part 2 : 1971 *Medium board, medium density fibreboard and hardboard.* Part 3 : 1972 *Insulating board (softboard).*

BS 1186 : Part 1 : 1986 *Timber for and workmanship in joinery. Part 1. Specification for timber.*

BS 1722 : Part 6 : 1986; Part 7 : 1986; Part 11 : 1986 *Specification for fences.*

BS 1990 : Part 1 : 1984 *Wood poles for overhead power and telecommunication lines. Part 1. Specification for softwood poles.*

BS 2548 : 1986 *Specification for wood wool for general packaging purposes.*

BS 2629 : Part 1 : 1967; Part 2 : 1970; Part 3 : 1978 *Specification for pallets for materials handling for through transit.*

BS 3470 : 1975 *Specification for field gates and posts.*

BS 3809 : 1971 *Specification for wood wool permanent formwork and infill units for reinforced concrete floors and roofs.*

BS 4072 : Part 1 : 1987 *Wood preservation by means of copper/chromium/arsenic compositions.*

BS 4471 : 1987 *Specification for sizes of sawn and processed softwood.*

BS 4978 : 1973 *Specification for timber grades for structural use.* (Revision expected late 1988 or early 1989.)

BS 5268 : Parts 1–7 *Structural use of timber.* (See p. 16 for individual titles.)

BS 5502 : 1978–86 *Code of practice for the design of buildings and structures for agriculture.*

BS 5589 : 1978 *Code of practice for preservation of timber.*

BS 5669 : 1979 *Specification for wood chipboard and methods of test for particle board.*

BS 5750 : 1987 *Quality systems.*

Other References

AARON, J. R. (1982). *Conifer bark: its properties and uses.* Forestry Commission Forest Record 110, 2nd edition. HMSO, London.

AARON, J.R. and OAKLEY, J.S. (1985). *The production of poles for electricity supply and telecommunications.* Forestry Commission Forest Record 128. HMSO, London.

ALDHOUS, J. R. and LOW, A. J. (1974). *The potential of western hemlock, western red cedar, grand fir and noble fir in Britain.* Forestry Commission Bulletin 49. HMSO, London.

BENHAM, C. A. (1986). Machine stress grading of British-grown larch. *Forestry and British Timber* 15 (2), 26–29.

BRAZIER, J. D., PRIEST, D. T., LAVERS, G. M. and WHITE, N. C. (1976). *An evaluation of home-grown Sitka spruce.* Building Research Establishment, Current Paper CP20.

BRAZIER, J. D., HANDS, R. G. and SEAL, D. T. (1985). The influence of planting spacing on structural wood yields from Sitka spruce. *Forestry and British Timber* 14 (9), 34–37.

BRITISH COAL (1984). *Coverboards of new hardwood and new softwood of British origin for use underground.* British Coal Specification No. 692.

BRITISH COAL (1985a). *Chock timber: new hardwood and new softwood of British origin for use underground.* British Coal Specification No. 666 (revised).

BRITISH COAL (1985b). *British softwood pitprops and split pitprops.* British Coal Specification No. 695 (revised).

BRITISH COAL (1986). *Specification for new softwood sleepers of British origin for use underground.* British Coal Specification No. 537 (revised).

BUILDING RESEARCH ESTABLISHMENT (1984). Specifying structural timber. *Building Research Establishment Digest* No. 287, July.

BUILDING RESEARCH ESTABLISHMENT (1986). *Timber drying manual* (2nd edition). Building Research Establishment Report BR 76.

CURRY, W. T. (1973). Machine stress grading of home-grown Sitka spruce. *Forestry and Home Grown Timber* **2** (1), 10–11.

DEPARTMENT OF TRANSPORT (1986). *Specification for highway works. Part 1* (6th edition). HMSO, London.

ELLIOTT, G. K. (1985). The potential for British timber within the UK market. Proceedings of the conference *The UK forestry industry – growing opportunities for profitable investment*, Institute for International Research, London.

ENDERSBY, H. J. and HANSFORD, A. C. (1971). The machine stress grading of home-grown Scots pine. *Timber Trades Journal, Forestry Supplement*, 30 October, 19–22.

FEWELL, A. R., BENHAM, C. A. and MOORE, G. L. (1982). How strong is British Douglas fir? *Forestry and British Timber* **11** (4), 15–17.

FOREST PRODUCTS RESEARCH LABORATORY. *Home-grown timbers series: Douglas fir* (1964); *Scots pine* (1965); *Larch* (1967a); *Sitka and Norway spruce* (1967b); *Lodgepole pine* (1968); *Corsican pine* (1972). (Now out of print.) HMSO, London.

FORESTRY COMMISSION (1986). *66th annual report and accounts 1985–86*, p. 8. HMSO, London.

TIMBER RESEARCH AND DEVELOPMENT ASSOCIATION (1984). Timber pallet boards. *TRADA Wood Information, Section 5, Sheet 2*.

TORY, J. R. (1978). Machine grading home-grown Corsican pine. *Forestry and British Timber* **7** (3), 36–37.

Printed in the United Kingdom for Her Majesty's Stationery Office.
Dd 289599 8/88 C50 3936 12521